1000 Text Messages

TXT TLK

Hw2 Tlk W/o
Bng Hrd

THIS IS A CARLTON BOOK

This edition published by Carlton Books Limited 2001
20 Mortimer Street, London W1T 3JW

A CIP catalogue for this book is available from the British Library.

ISBN 1 84222 424 7

Printed in Italy

Editorial Manager: Venetia Penfold
Art Director: Penny Stock
Project Editor: Zia Mattocks
Text and Design: Terry Burrows
Production Manager: Garry Lewis

text talk

1000 Text Messages

TXT TLK

Hw2 Tlk W/o Bng Hrd

how to talk without being heard

BEFORE YOU START

Sending a text message is as easy as making a phone call. In fact, it is a phone call, but you don't need to talk. Here are a few things you might want to check out before you get down to business.

Although specific operating instructions for sending text messages differ between manufacturers, they all work in roughly the same way. Before you start, however, it's a good idea to have a quick look through your phone or pager's user manual. Generally speaking, you'll need to start off by finding some kind of WRITE MESSAGE command. You then simply tap out the text using the phone keypad. Each of the number pads also correspond to three or four different letters: the "2" key, for example, can also be used to generate the letters "A," "B," and "C." If you want "A" you press the pad once; for "B" press it twice; for "C" press it three times. When you're happy with the text, type in the phone or pager number of the person you want to receive it, choose a SEND MESSAGE command and whoosh, it's on its way. Of course, it can only be received by a phone that has text facilities!

If all this sounds a bit complicated, don't worry. It's a whole lot easier than it sounds!

contents

text crazy

your cool guide to get you started on the ride

It's funny to think that ten years ago hardly anyone had a cell phone. Those who did were rich, famous and had MASSIVE biceps – some of those things were the size of a small refrigerator! Now, of course, EVERYONE has a cell phone or a pager of some description. We've come to depend on them. Where would most of us be without one? Sad, lonely, and lost, that's where.

But you can do a whole lot more with a cell phone than just talk. A few years ago

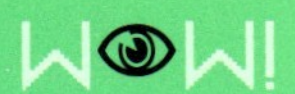

Text messaging has become so popular that a staggering 15 BILLION were sent across the globe during the first four weeks of 2001. By 2004, the worldwide market for mobile communication gadgets is expected to exceed 60 billion.

some clever geeks came up with the brainy idea of letting cell phones transmit little messages to one another – just like the ones used by paging devices. In most cases, they thought, this great new facility would be used for those rather boring suit-type things like setting alarm clocks or arranging meetings. Little did they know they were about to unleash nothing less that a fully fledged communications revolution.

Whether it was called SMS, GSM, G-mail or plain old TEXT MESSAGING, it quickly became a phenomenally popular and rather cool way of communicating, especially among the under-25 age group. In fact, over 500 million people all over the world now regularly send and receive text messages.

so why have we gone text crazy?

Glad you asked. There are loads of reasons. Maybe it's because text messaging has all the regular advantages of a cell phone without turning you into an embarrassing public menace. Let's be honest, nobody likes to be stuck on a train next to someone yelling loudly into their phone. It's just a drag. Text messaging, on the other hand, is discreet. A quiet bleep or gentle, sexy vibration from your phone or pager is all that's needed to let you know that a message has arrived.

Text messaging is also cheaper than making a regular call: each message costs only a few cents. And when have you ever managed

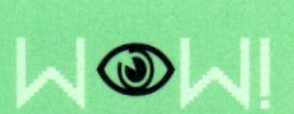

Not everyone is a huge fan of text messaging. Indeed, there are some academics out there who fear that it could even affect levels of literacy among future generations. They claim that it may restrict people's communication skills, which could ultimately affect the way they talk to each other. In practical terms, though, as long the message gets through loud and clear, does it really matter?

to phone a friend to arrange a meeting place without it turning into a half-hour chat on the latest hot gossip going around?

OK, enough with the sensible reasons. Mostly we text because it's FUN, and it can bring us closer to our folks and friends. But it can also be practical fun. Sometimes it's easier to talk without speaking. Feeling tongue-tied or embarrassed when you want to tell your one-and-only how you feel? Texting over a discreet ILuVU should do the trick. Got a crush and a number but not sure how to take things a stage further? Why not try something as simple as DUWnt2GoOut2nite? – who could resist an offer like that?

Even the most expensive cell phones have only a limited room to store messages. When it gets filled up you may not be able to receive new messages until you have made space. Therefore it's a good idea to get into the habit of deleting your unwanted messages.

how does it all work?

Like the Internet, text messaging has evolved it's own system of language and grammar. Indeed, some of it comes directly from common e-mail shortcuts.

You can take a combination of three basic approaches to sending your own text messages. Let's call them ACRONYMS, VOWEL-DROPS and EMOTICONS.

acronyms

An ACRONYM is a made-up word where each letter represents the first letter of a another word. For example, PCM can be used as a shorthand for PLEASE CALL ME. Similarly, HAND can be used simply to mean HAVE A NICE DAY. Acronymns such as these can save you loads of time (and money if your messages are charged by the character). HOWEVER, if you use acronyms you MUST make sure that they are common currency: it's no good texting

HAASATAPATMWCGTTM?

if your buddy doesn't know that you mean HOW ABOUT A SODA AND THEN A PIZZA AND THEN MAYBE WE CAN GO TO THE MALL?

vowel-dropping and phonetics

VOWEL-DROPPING is a safer way to text. No surprises here, this means leaving out most of the vowels from your messages. For added clarity, it's a good idea to start each new word with a capital letter. Some texters also like to leave a space between words, but although this can clarify meaning it also takes up extra time and space. As ever, of course, the choice is yours – nobody's going to give you an F in the world of text messaging. (Well they might, but that's an altogether different story!)

Another popular shortcut is to play with phonetics – the way letters and numbers sound. For example, instead of writing out the word

FOR, you can use the number 4; instead of texting the word YOU, the letter U will do nicely. In this way, the phrase LET'S GET TOGETHER can be neatly abbreviated to LtsGt2gthr. You can take this approach further by shortening key words so that, for example, LOVE is transformed into Luv.

Another common practice is to indicate double letters within a word as a capital letter. In this way, SORRY might be shown as SRy. You get the picture, right? We'll be using this approach throughout the book: although it needs careful interpretation, most people will understand what it means, even if they choose not to use it themselves.

Emoticons

The little faces created from punctuation marks emerged on the Internet as a way of conveying emotion in text. Typed words can be misleading on their own. Are you being funny, sarcastic, or downright mean? It's sometimes hard to tell. You can of course emphasize words by TYPING THEM IN CAPITALS but that's likely to be interpreted as shouting,

Emoticon symbols maybe creative and fun to use when you send e-mails, but on cell phones and pagers they can also be a bit of a pain. Unlike computer keyboards, mobile gadgets don't have enough pads for all of the punctuation marks. Instead they are all contained within a special screen that you have to "click" your way through.

HOT TIPS

The first and last rules of text messaging are simple: THERE ARE NO RULES. Texting is all about effective communication. As long as the person receiving your message can understand your acronyms and abbreviations then anything you choose to send is good.

which may be a little rude. A slightly more subtle alternative is to create little faces using the punctuation marks found on your phone or pager's keypad. The most famous emoticon is the SMILEY face. You can create this by simply using a colon, dash and closed brackets – just like this :–). If you then turn the page around 90 degrees clockwise you'll see the effect.

Of course, there are numerous other possibilities. For a great text-messaging game why not send mystery emoticons to your friends and let them figure out their meaning? Or you could try out the FAMOUS NAMES GAME. It's really easy and lots of fun.

Question: Who is this?

@@@@:-)

Answer: Marge Simpson

So there it is. Welcome to the wild and wacky world of text messaging. It can be as serious, stupid, creative, practical or as fun as you want it to be. As with everything else, the choice is yours. So what are you waiting for? LtsGtTxtng.

just press the buttons

the basic tools you need to get yourself up and running

Getting to grips with text messaging is just like learning any other new language. So, you clearly need to equip yourself with the basics before you can get down to any SERIOUS BUSINESS. So let's start out with a list of really simple expressions. The acronyms and phonetic abbreviations you'll find over the next few pages are the kind of thing that everybody who texts should know.

HOT TIPS

You might find text messaging a bit weird and confusing at first. Don't worry about it. If you stick with it you'll discover that the more you send and receive messages, the easier it gets.

2	To/two/too
2day	Today
2moro	Tomorrow
4	For
4WIW	For what it's worth
@	At
AAMOF	As a matter of fact
AFAIC	As far as I'm concerned
AFAIK	As far as I know
AML	All my love
AKA	Also known as
ASAP	As soon as possible

ATB — All the best

ATM — At the moment

ATT — About time too

B — Be

BBFN — Bye bye for now

BBL — Be back later

BBS — Be back soon

B/C — Because

BCNU — Be seeing you

B4 — Before

BFN — Bye for now

BRB — Be right back

BS — Bull shit

BTW — By the way

Bwd — Backward

BYF — Bring your friends

BYKT — But you knew that

C — See

HOT TIPS

Don't forget that you can combine the sounds of letters and numbers to create new words. For example, GREAT can be texted as Gr8. If you're lucky you can also give yourself a cool phonetic identity – if your name is Kate, you can sign off your messages as K8.

C&G	Chuckle and grin
CMIIW	Correct me if I'm wrong
CSG	Chuckle, snigger, grin
CU	See you
CUL8r	See you later
CYA	See you (see ya)
DLTBBB	Don't let the bed bugs bite
Doin	Doing
EOL	End of lecture
Esp	Especially
FAQ	Frequently asked question

FITB	Fill in the blank
F2T?	Free to talk?
FUBB	F***** up beyond belief
Fwd	Forward
FYI	For your information
GFN	Gone for now
GG	Good game
GMBO	Giggling my butt off
GMD	Get my drift
GMTA	Great minds think alike

Much to the annoyance of unpopular President Joseph Estrada, the people of the Philippines have taken to texting in a VERY big way. In fact, he believes that rumors and rude jokes about him are being circulated by rebel forces in an effort to destabilize his administration. Just do a web search for "text jokes" and you'll find plenty of Philippino sites devoted to the mockery of their ruthless ruler!

GoNa Going to (gonna)

Gr8 Great

GSOH Good sense of humor

HAGN Have a good night

H&K Hug and kiss

HAND Have a nice day

HHIS Hangs head in shame

HTH Hope this helps

HUH? Have you heard?

H2 How to

H8 Hate

IAC In any case

IAE In any event

IC I see

IDK I don't know

IIRC If I remember correctly

IK I know

IKWIM I know what I mean

IKWYM	I know what you mean
IMHO	In my humble opinion
IMNSHO	In my not so humble opinion
IOW	In other words
IRL	In real life
ISH	I'm still here
ISU	It's so unfair
IYSS	If you say so
JK	Just kidding
JMO	Just my opinion
L8	Late
L8r	Later
LF	Looking fit
LHU	Lord help us!
LMK	Let me know
LOL	Laugh out loud
Luv	Love
MMD	Make my day

MYOB	Mind your own business
NE	Any
NE1	Anyone
Neone	Anyone
Nevr	Never
NTMA	Nobody tells me anything
No1	Number one
OIC	Oh, I see
POAHF	Put on a happy face
PLS	Please
Ppl	People
PM	Private message
R	Are

HOT TIPS

Many of the acronymns, abbreviations and phrases have come straight from the evolution of e-mail. So, if in doubt, think Internet!

Re	Regarding
ROFL	Rolls on floor laughing
RU?	Are you?
SE2E	Smiling ear-to-ear
Spk	Speak
Sry	Sorry
SWL	Screams with laughter
TA	Thanks again
TAB	Thanks a bunch
TFN	Thanks for nothing
Tho	Though
ThnQ	Thank you
Thru	Through
Thx	Thanks
TNX	Thanks
U	You
U@?	(Where are) you at?
UOK?	(Are) you OK?

UR	You are
Usu	Usually
W/	With
Wan2	Want to
WB	Welcome back
WerRU	Where are you
WHM?	What's happening, man?
W/O	Without
W8	Wait
W8ng	Waiting
WRT	With respect to
YBS	You'll be sorry
YFMO	You freak me out
YG	Young gentleman
YL	Young lady
YM	You mean
YR	Yeah, right!
YSW	Yeah, sure, whatever

drop them vowels

Using acronymns relies completely on the receiver knowing the precise meaning of the abbreviation.. Let's instead look at a few phrases that involve dropping vowels. Once you get familiar with the rules you should be able to figure out the meaning. To get you started, let's take a look at an example.

DdUHrtYrslfWenUFelFrmHvn?

If you begin by thinking of each capital letter as a separate word, trying to pronounce each word should make the meaning clear:

Dd=did	U=you	Hrt=hurt
Yrslf=yourself	Wen=when	U=you
Fel=fell	Frm=from	Hvn=heaven

WATCH OUT!

Don't forget that a capital letter may not only indicate the start of a new word, but can also be used to replace a double letter.

ARIHny
A real honey

BBILuvU
Baby I love you

Bstd
Busted

CFICr
See if I care

CntWtNuthrMinit
Can't wait another minute

DmpHm
Dump him

DmpHr
Dump her

DntHavACwMn
Don't have a cow, man

DntItMkUFLGd?
Don't it make you feel good?

GtYaTkts?
Got your tickets?

HwCOlIsTht?
How cool is that?

HwMch?

How much?

HwsItHngng?

How's it hanging?

IfThsAntLuvIDntNoWtIs

If this ain't love, I don't know what is

ItsAL2Mch

It's all too much

LknGd

Looking good

LtsGt2gthr

Let's get together

LtsRck&RL

Let's rock and roll

LvLng&Prspr

Live long and prosper

RntUMyBlndDt42nite?

Aren't you my blind date for tonight?

MaTFrcBWU

May the force be with you

NtMyFlt

Not my fault

Rdy2RL?

Ready to roll?

ShsARlMPt

She's a real muppet

ShsASlyr

She's a slayer

ShsATtlBbe

She's a total babe

ShsTtlyBo

She's totally bo

TngDrtbg

Teenage dirtbag

ThtBlwMeAwy

That blew me away

ThtsOFTHOk

That's off the hook

ThtsWck

That's wack

ThtsSoLm

That's so lame

ThtTrcksRnsn

That track's rinsin'

VryFn

Very fine

Wnt2GtADrnk?

Want to get a drink?

WtSGoinOn2nite?

What's going on tonight?

YrPrtyFly4AWhtGy

You're pretty fly for a white guy

show your feelings

Now let's take a look at a handful of the most commonly used emoticons. These can be used to tell people exactly how you feel, either in terms of your mood or perhaps as a curt response to a previous message. The basic idea is pretty simple. In the original smiley face, a colon is used represent eyes, a dash is the nose, and a closed bracket is the smiling mouth. You can see this in the example below.

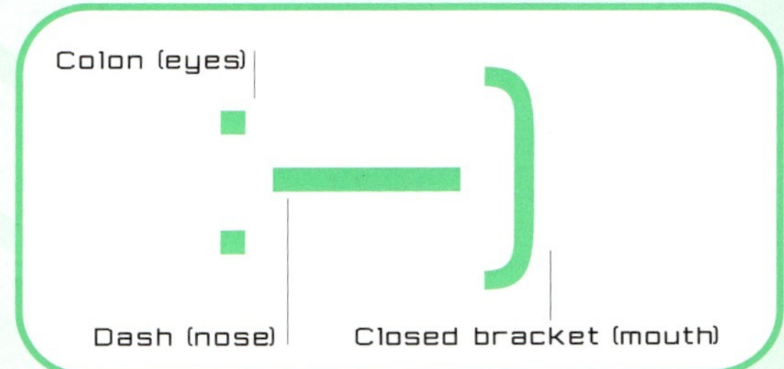

However, other letters or punctuation marks can be added to produce different effects. For example, an open bracket can be a sad, drooping mouth; a semi-colon a winking eye; lower-case letter d a baseball cap. You really are limited only by your imagination.

:—)	Happy
:—))	Very happy
:—(	Sad
:—((	Very sad
:—||	Angry
:'—(	Crying
%—\	Confused
:—>	Devilish grin
;—>	Devilish wink
[:—(	Frown

COOL STUFF

Use emoticons to share your mood or provide a succinct reaction to a message. The question **UOK?** ***(Are you alright?) could perhaps be answered with*** **:-[** ***(I'm really sad) or*** **:-|** ***(I'm not talking to you).***

:-/	Frustrated
:-*	Blowing a kiss
:-D	Loud laugh
:-|	Not talking
:O	Loud yell
:-@	Screaming
:-o	Shocked
:'-O	Shocked and upset
:-p	Tongue-in-cheek
d:-)	Hats off
[:-(	Frowning
:-[	Feeling down

:->>	Huge grin
:-))))	REALLY happy
{}	No comment
:-&	Tongue-tied
:-X	Big wet kiss
M:-)	I salute you
:-#	Lips are sealed
(:-...	Heartbroken
:-e	Disappointed
:-t	Upset and pouting
;(	Cheer up
;)	Smirk
:l	Hmmm ...
:}	What?!
8-)	Wearing glasses
B-)	Wearing shades
:c	Very unhappy

:Y	Whispered aside
;?	Wry remark
:?	Licking lips
:-9	Licking lips
:~)	Yum, yum
:'')	A bit embarrassed
X-(	Mad with rage
8O	Oh my god!

WOW!

The first emoticons appeared in the early days of the Internet. Although difficult to prove conclusively, the earliest record seems to be from April 12 1979, when a guy named Kevin MacKenzie mailed a newsgroup caliing for ways of emotion to be incorporated into the dry text medium. He suggested that -) could be used to indicate a sentence was tongue-in-cheek. His mail was heavily flamed by the geeks who were on-line at that time, but gradually caught on as more and more non-acedemics gained access to the Net.

M:-O	Respect
8-]	Like, wow!
P-)	This is a come-on
+<#^V	I'm your knight in shining armor
|-|	Bored to sleep
:-p?	Wassup?
;[]?	Feeling hungry?
:-|+	Not amused
~~:-(	Smokin' with rage
;-((	Faking anger
:-...E]	Crocodile tears
:-(B<-	In a REALmess
<:*(	Embarrassed at my own stupidity
:-(~)	Sticks tongue out
@;-)	Flirt

love talk

whether flirting, getting a hot date or sending a love note to someone special, who needs flowers when you've got text?

Text and romance were just made for one another. And it's especially cool if you're a little on the shy side. Texting your feelings can let you be as wry, clever, sassy, sexy, mysterious, or in-your-face as you want.

TCBYLD
This could be your lucky day

TCBYLN
This could be your lucky night

UHvABF?
Do you have a boyfriend?

UHvAGF?
Do you have a girlfriend?

HOT TIPS

There are loads of romantic symbols that you can use when texting love matters. A KISS is represented by the traditional X, (or *X*) and a HUG is *H*. A commonly used emoticon for HUGS AND KISSES is (()):**. The most common symbol used to represent the HEART is <3 (it's actually supposed to be a beating heart).

GW?

Guess who?

WGYMN?

Who gave you my number?

>—::—D

I've been struck by cupid's arrow

:—6

Losing sleep over you, baby

@}>—,—'——

A rose for you

CnIF1rtWU?

Can I flirt with you?

(*_*)

Goggle-eyed for you

UWan2Gt2gthr?

Do you want to get together?

RUSmart?

A mysterious message arrives. All it says is:* HAY&M? *What do you make of it?

a) How are you and Mom?

b) Do you have any yams and mangos?

c) How about you and me?

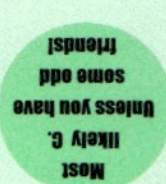

IAD

It's a date

ImWUHBW8tng4

I'm what you have been waiting for

INoSumwerC1WeC1dMt

I know somewhere cool we could meet

ULkALtLkMyNxtBF

You look a lot like my next boyfriend

IfHeDsntShwUpImHre

If he doesn't show up, I'm here

HrsYaChnc2Gt2NoMe

Here's your chance to get to know me

DntUNoMeFrmSumwer?

Don't you know me from somewhere?

SOSImLstWhchWy2YrPlce?

Help, I'm lost – which way to your place?

RURdy2GoHmeNw?

Are you ready to go home now?

MaINdThsSntnceWAPropstn?

May I end this sentence with a proposition?

ImLOkng4AFrndDoUWan2B MyFrnd?

I'm looking for a friend. Do you want to be my friend?

IfISedUHdAButf1BdyW1dUH1d ItAgnstMe?

If I said you had a beautiful body would you hold it against me?

DoUMndIfIFntsizAbtU?

Do you mind if I fantasize about you?

W1dULkSum12MxWYrDrnk?

Would you like someone to mix with your drink?

XcusMeCnUGivMeDrctns2Yr <3?

Excuse me, can you give me directions to your heart?

ICntSOSFLngInLuvWU

I can't help falling in love with you

you could be anyone

Of course, when you send a message, t receiver may have no idea who you a Sometimes it's nice to give a clue, other tim an air of mystery is more fun. And sometim you might want to laugh.

:-	I'm a man
:-)8 -	I'm a man
>-	I'm a woman
:-)3>-	I'm a woman
O:-)	I'm a smiling angel
O;-)	I'm a winking angel
((Y))	I'm a fat lady
:-[	I'm a vampire
-:-)	I'm a punk rocker

+:-)	I'm the Pope
{:<>	I'm Donald Duck
:-)#	I'm a man with a beard
:-)##	I'm a man with a long beard
:-)3#	I'm a woman with a beard
:-#	I'm a man with a moustache
:-)J	I'm a surfer
:-)o->	I'm a suit
:-.)	I'm Madonna
:-.)	I'm Marilyn Monroe
:-+	I wear too much lipstick
:---)	I'm a liar (Pinocchio)
:-----)	I'm a BIG liar
:-[x>	I'm Count Dracula
::-)	I wear glasses
:-~)	I'm someone with a cold

:::-)	I'm the Elephant Man
:>)	I have a big nose
:@)	I have a pig face
:-{}	I love lipstick
:-)>	I love a man with a goatee
D)	I'm wearing a crash helmet
<:-))	I'm stupid but happy
<:-(	I'm a sad dunce
=:X	I'm a bunny
@ -)	I'm a happy cyclops
?:-)	I'm got wavy hair
@:-)	I wear a turban
%-/	I've got a hangover
%%-/	Really bad (love) hangover
$:)$	I'm feeling rich
#@#:-(	Bad-hair day

%:-))) Drunk with joy

@@:-{} Bouffant babe

++<3++ I'm infatuated

<3 I'm a model

<&-| I feel like an idiot

<*)))-{ Catch me if you can

*8-| You're a geek

...---... Heeelllppp!

///:-) Love the new haircut

2B|^2B ... that is the question

8-(*) Just ate a jalapeno

:-'X I have a bad cold

;-)8 Sly wink from a big girl

;'-D Crying with happiness

:-)~(-: The two of us kissing

;-{}3 Vamp

the dates get hotter

You know you're smitten, but how do you move things up a notch without getting tongue-tied or plain embarrassed? Don't say a word – just trust the text.

GF	Girlfriend
BF	Boyfriend
ULkMe?	Do you like me?
ILkU	I like you
:–)(–: +	Feel like a kiss?
GetYaCotUveGtMe	Get your coat, you've got me!
HwDoULkMeSoFar?	How do you like me so far?
SoHwMIDOin?	So ... how am I doing?
OKYLDo	OK, you'll do

PdnMeIsUrStTkn?

Pardon me, is your seat taken?

StndStLIWnt2PckUUp

Stand still, I want to pick you up

DUBlvNFate?

Do you believe in fate?

IveCum2StElUr<3

I've come to steal your heart

IsItHtNHrOrIsItJstU?

Is it hot in here or is it just you?

IDdntNoAnglsFlwSoLo

I didn't know angels flew so low.

UrPlceOrMne?

Your place or mine?

WrHvUBEnALMyLfe?

Where have you been all my life?

UCnDoMgc

You can do magic

BBkSnSwthrt

Be back soon sweetheart

CSThnknAbtU

Can't stop thinking about you

UBlwMyMnd

You blow my mind

StckOnU

Stuck on you

FlLkSwPngSpt?

Feel like swapping spit?

nothin' much to say

Ah, when love is new ... there's nothing like it. A ringing phone, a mailman's knock, a message bleep and the heart starts racing. But it's really all about getting to know someone, which a lot of the time can just mean chatting away about any old stuff ... or nothing really at all.

1dafL	Wonderful
3sum	Threesome
AWHFY?	Are we having fun yet?
Bf	Boyfriend
BMB	Be my baby
BML	Be my love
BMS	Be my sweetheart
BTTP	Back to the point
CB	Call back

WOW!

On Valentine's Day 2001 vacationer Rebecca Fyfe and 17 others were stranded in a boat off the coast of Indonesia after their engine had failed. They were only rescued after she had texted a message to her boyfriend, Nick Hodgson. Back in England, he was able to alert coastguards in Southeast Asia. You never know, one day a text message might just save your life!

CHUR — See how you are

CUBL8r — Call you back later

DoUThk? — Do you think?

DK — Don't know

DTRT — Do the right thing

DUCWIC? — Do you see what I see?

FFL — Fool for love

FncyAShg? — Fancy a shag?

GA	Go ahead
Gf	Girlfriend
GMTA	Great minds think alike
GTBOS	Glad to be of service
H&K	Hugs and kisses
HCAY	Hugs comin' at ya
HH	Holding hands
HOEW	Hanging on every word
HOHIL	Head over heels in love
HSIK?	How should I know?

HOT TIPS

Overcome with romantic thoughts while sitting at your desk? Everyone loves flowers, so why not send a lovely rose? Or even a dozen?

@}>-'-,--

12X---<--@

HTEI	Hope this explains it
HTH	Hope this helps
IDGI	I don't get it
IDTS	I don't think so
IKT	I know that
ILU	I love you
ILY	I love you
INT	I'll never tell
IOU	I owe you
IOU 1	I owe you one
IOYWH	If only you were here
IWIK	I wish I knew
IXU	I love you
IXXXXU	I love you lots
KOTL	Kiss on the lips
KWIM?	Know what I mean?

Hw2MprSAWmn: LuvHr,CmfrtHr,XXXHr&RspctHr.

How to impress a woman: Love her, comfort her, kiss her and respect her.

Hw2MprSAMan:TrnUpNkd.BrngBEr.

How to impress a man: Turn up naked. Bring beer.

HK	Hot kiss
IWBNI	It would be nice if ...
J1MX	Just one more kiss
JAM	Just a minute
JAS	Just a second
LBF	Let's be friends
LJBF	Let's just be friends
LMT	Love me tonight
LTNH	Long time no hear
LTNC	Long time no see
LOL	Lots of love

LOL Laughs out loud

LULAB Love you like a brother

LULAF Love you like a father

LULAM Love you like a mother

LULAS Love you like a sister

Luv Love

MGB May God bless you

M<3IY My heart is yours

NAGI Not a good idea

NIAA No idea at all

NQA No questions asked

OL Old lady

OM Old man

PDA Public display of affection

PDC Pretty damn cute

PDQ Pretty damn quick

RFK	Request for kiss
RUF2T?	Are you free to talk?
RUOK?	Are you OK?
SSD	Signed, sealed, delivered ...
SVS	Someone very special
SWAK	Sealed with a kiss
SxyBy	Sexy boy
SxyGrl	Sexy girl
TC	Take care
TCB	Taking care of business
TCL	Take care, love

Q: WtWldTWrldBLkW/oMn?

Question: What would the world be like without men?

A: FLOvFtHPyWmn

Answer: Full of fat, happy women

TFLN...	Thanks for last night ...
...&2moro	... and tomorrow
TGIL	This guy/girl's in love
TLC	Tender loving care
TLLU	The lady loves you
TOFU	This one's for you
UBsy?	Are you busy?
URE2M	You are everything to me
URHS	You are hot stuff
URVS	You are very special
URVSxy	You are very sexy
UR1dafl	You are wonderful
VS	Very special
WAB	What a babe!
WerUBin?	Where have you been?
Wot?	What?

just the two of us

Where words of love are concerned, texting is just the best. And everyone's a winner: Now you no longer have to gross out your friends and family with cute talk and pet names. It's a virtual universe just made for two.

BK
Big kiss

F2F
Face to face

HITULTILuvU?
Have I told you lately that I love you?

IC**WenUXMe
I see stars when you kiss me

ILUM&MED
I love you more and more each day

JTM
Je t'aime

ILUWAM<3
I love you with all my heart

URAQT
You are a cutie

SWALK

Sent with a loving kiss

TMIY

Take me I'm yours

URMyEvrythng

You are my everything

WLUStLLuvMe2moro?

Will you still love me tomorrow?

GBH

Great big hug

GBH&K

Great big hug and kiss

GBH&KCB

Great big hug and kiss coming back at you

K

Kiss on the cheek

XXXXXXXX

Very big (and long) kiss

KB

Kiss back

KOTC

Kiss on the cheek

LtsHkUp

Let's hook up

MAY

Mad about you

VH

Virtual hug

ImADctd2U

I'm addicted to you

Alwys&4evr

Always and forever

ICntGtEnufOfUBaB

I can't get enough of you, baby

CntW82CU2nite

Can't wait to see you tonight

4evrInLuv

Forever in love

H1dMeC1sBaB

Hold me close, baby

Hp1S1yDvotd2U

Hopelessly devoted to you

IDntDsrvU...

I don't deserve you ...

...OrMaBIDo

... or maybe I do

StaWMe2nite

Stay with me tonight

IgotUBbe

I got you babe

ImT14U

I'm the one for you

MakThsANite2Rmbr

Make this a night to remember

SoInLuvWU

So in love with you

HtMeBb1MorTme
Hit me baby one more time

EvryLt1ThngUDIsMgc
Every little thing you do is magic

ALWeNEdIsLuv
All we need is love

HwAbtU&IGtOutofThs WetC1ths?
How about you and I get out of these wet clothes?

BeMyV1ntn4evr
Be my valentine ... forever

DUHvPrtctn?
Do you have protection?

My<3IsYrs
My heart is yours

LtMeLckU
Let me lick you

RUSmart?

You get stood up, and so decide to put your feelings into text. Which would be the most appropriate one?

a) DoThtAgn&YrDed

b) YrDmpd!

C) :'-(

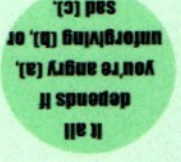

smooth talkers

Over the centuries, poets, writers and composers have had their say on the great matter of Love.

HwDoILuvThE?LtMe CntTWys

How do I love thee? Let me count the ways [Browning]

UShdHvASftrPLwThnMy<3

You should have a softer pillow than my heart [Byron]

LuvConqrsAL&We2 SCmb2Luv

Love conquers all, and we too succumb to love [Virgil]

4Luv&Buty&DliteTher IsNoDethNrChnge

For love, and beauty, and delight, there is no death nor change [Shelley]

ShLICmprThE2ASMrsDy? ThouRtMrLvly&MrTmprt

Shall I compare thee to a summer's day? Thou art more lovely and more temperate [Shakespeare]

2CHrWs2LuvHr,LuvBt Hr&Luv4evr

To see her was to love her, love but her, and love forever [Burns]

Bf: MIYr1st?

Boyfriend: Am I your first?

Gf: UCldB,ULkFmlr

Girlfriend: You could be, you look familiar.

AThngOfButyIsAJy 4evrItsLuvlinSIncrsIt WLNvrPSN2OthngnS

A thing of beauty is a joy forever: It's loveliness increases; it will never pass into nothingness. [Keats]

URAlwysNuTLstOfYr XXXWsEvrTSwEtst

You are always new. The last of your kisses was ever the sweetest. [Keats]

WeWer2&HdBt1<3

We were two and had but one heart [Villon]

OThoArtFrrThnTEvngAir CldInTButyOvAThsndStrs

Oh, thou art fairer than the evening air, clad in the beauty of a thousand stars [Marlowe]

getting together

top tips for talking time

Whether you're meeting up with your buddies or getting it together with someone special, texting is the coolest way to get your night off to a smooth start.

2day	Today
2moro	Tomorrow
2sdy	Tuesday
ADN	Any day now
ASAP	As soon as possible
ATM	At the moment
ATUL	Any time you like
BTOBS	Be there or be square

If you use text messaging for getting dates or meeting people, exercise caution, especially where relative strangers are involved. Keep your initial meetings to public places and, if possible, let a friend or member of your family know your plans. Better safe than sorry.

CU2nite@8	See you tonight at eight
DUWnt2GoOut?	Do you want to go out?
ETA	Estimated time of arrival
GnaMakUAnOFrUCntRfs	Gonna make you an offer you can't refuse
ILBThr4U	I'll be there for you
ItsNwOrNvr	It's now or never
KIT	Keep in touch
LEK	Let everyone know

LMK	Let me know
MED	Immediately
NEdU2nite	Need you tonight
NEd2CU2nite	Need to see you tonight
NON	Now or never

HOT TIPS

You can save yourself time and money with some phones and pagers by simultaneously sending out messages to a group of people. This feature may have come about as a way for managers to keep in touch with on-the-road sales staff, but it was quickly comandeered for its recreational potential. Although systems vary from between companies, it's usually pretty easy to set up. You simply give your "group" a name and link it to all of their numbers. If you want to contact everyone you only have to send the message once. For those with a wide circle of friends, it's usually possible to set up a number of different groups.

PIO — Pass it on

PlsDntGo — Please don't go

Sec — Second

SIPUU? — Shall I pick you up?

ASA — As soon as

STSP — Same time, same place

TBA — To be announced

TBC — To be confirmed

UP? — Usual place?

U+Me=Luv — You + Me = LOVE

W84Me@ — Wait for me at …

WenDoWeM? — When do we meet?

WerRU? — Where are you?

WSWM? — Where shall we meet?

MyPlcOrYrs? — My place or yours?

signing off

When you gotta go, you just gotta go ...

AMA	Adieu, mon amie
AMC	Adieu, mon cherie
ATB	All the best
AWS	Auf Wieder Sehen
BBIAB	Be back in bit
BF	Best friends ...
CB	Ciao, baby
CBS	Call back soon
CU	See you
CUWUL8r	Catch up with you later
Cya	See you
CYL	See you later
CW	Can't wait
CWYL	Chat with you later

Q:YWntUMndYrOwnBsnS?

A:UveGtNoMnd&GtNoBsnS

Question: Why won't you mind own own business?

Answer: You've got no mind and got no business.

G2G	Got to go
HLVB	Hasta la vista, baby
ICD	I'll count the days
ICS	I'll count the seconds
IHMB	I'll hold my breath
IMD	I must depart
IOH	I'm outa here
JG	Just groovin'
L8rB	Later, baby
LWU	Lost without you
ML	More later

NN	Night-night
O&O	Over and out
PO	Peace out
Spk2UL8r	Speak to you later
Synra	Sayonara
TAF	That's all folks
TNT	Till next time
TTYL	Talk to you later
T2UL8r	Talk to you later
UT	Until then

HOT TIPS

Feel like coming over all retro? How about a touch of 1950s hip-talk? Taken from the words of a famous rock and roll song, instead of saying goodbye, one person would say "Alligator," the other would reply "Crocodile". (It went: "See you later Alligator/After a while crocodile." Oh, ask your folks to explain ...)

GdntGdntPrtngIsSchSwtSRw ThtIShLSyGdntT1ltBMRw

Goodnight! Goodnight! Parting is such sweet sorrow. That I shall say goodnight till it be morrow. [Shakespeare]

IHavALngJrny2Tk&MstBd TCoFrwL

I have a long journey to take, and must bid the company farewell. [Raleigh]

TsBTr2BLftThnNevaHavBn Luvd

Tis better to be left than never have been loved. [Congreve]

IfWeDoNtMtAgnYWeShLSm1 IfNtYThnThsPrtngWsWLMd

If we do not meet again, why, we shall smile! If not, why then, this parting was well made. [Shakespeare]

CnIEvrRdThsJysFrwL?

Can I ever rid these joys farewell? [Keats]

Swts2TSwtFrwL

Sweets to the sweet, farewell! [Shakespeare]

hotheads

Had a break up with your partner? Are your friends being a pain in the ass? Got something you want to get off your chest? Don't waste another second – get ready with the flaming text!

IWdntBSDWU
I wouldn't be seen dead with you

IWMHT
I'm washing my hair tonight

IWMHT&EON
I'm washing my hair tonight and every other night

ILCUInMyDrms...IMn Ntmrs
I'll see you in my dreams ... I mean nightmares!

ItsAShmeUrPrntsDdnt PrctsSafSx
It's a shame your parents didn't practice safe sex

HOT TIPS

Some of us have a crisis of confidence when dealing with face-to-face confrontation. That shouldn't be the case with text. Don't be afraid to clear the air – just go for it!

NIYWTLPITW
Not if you were the last person in the world

SIJLIAC
Sorry, I just lapsed into a coma

PYMIGBOYM
Put your mind in gear before opening your mouth

RUKDng?
Are you kidding?

UrPrOfThtICnTkAJk
You're proof that I can take a joke

WTHDTTM?
What the hell does that mean?

SUNo1ILTY
Shut up, no one is listening to you

DoSmSoulSrchng.U MghtJstFnd1
Do some soul searching. You might just find one

TYVL,YWEL

Thank you very little, you're welcome even less

MnRLkLefnts.ILk2Lk@ThmBtWdntWnt2Own1

Men are like elephants. I like to look at them but wouldn't want to own one

URLkACrAlrm.UMkALtOfNseBtNo1PysAnyTntn

You're like a car alarm. You make a lot of noise but no one pays any attention

URSchALsr

You are such a loser

HwSdIsTht?

How sad is that?

YDntUGo&PlWYrslf?

Why don't you go and play with yourself?

IveHdAWndrflEvngBtThsWsntIt

I've had a wonderful evening, but this wasn't it

IsThrNoBgnng2YrTlnts?

Is there no beginning to your talents?

USck

You suck

UHvEvryRght2BUgly BtImAfrdUveAbusdTht Privlge

You have every right to be ugly, but I'm afraid you've abused that privilege

IDntKnwWhtMksUSo DmbBtItWrks

I don't know what makes you so dumb, but it works

StDwn&GivYrMndA RstItObvslyNEdsIt

Sit down and give your mind a rest – it obviously needs it

RUAlwysThsDmbOrRU MkngASpclFrt2day?

Are you always this dumb or are you making a special effort today?

BrnsRntEvrythng&In YrCseThrNthng

Brains aren't everything – and in your case they're nothing

DntLtYrMndWndrIts2 SmL2BOutOnItsOwn

Don't let your mind wander. It's too small to be out on it's own

UStnk

You stink

IfUWerTwceAsSmrtAs URNwYoudStLBStpd

If you were twice as smart as you are now you'd still be stupid

IDntMndUTlkngSoLng AsUDntMndMeNtLstng

I don't mind you talking, as long as you don't mind me not listening

BgTmLsr

Big-time loser

ImBsyNow.CnIIgnrU SmOthrTme?

I'm busy now. Can I ignore you some other time?

URmndMeOvTC.UMkMe Sck

You remind me of the sea. You make me sick

ICnTLUrLyngUrLpsMvd

I can tell you're lying – your lips moved

UvGtMorFcsThnMnt Rshmr

You've got more faces than Mount Rushmore

UrNtYrslf2dy.INtcdT MprvmntRghtAwy

You're not yourself today. I noticed the improvement right away

in short

When words fail, you need something quick and simple – better make it smart, though!

AYM?	Are you mental?
CFD	Call for discussion
CIfICr	See if I care
CIO	Cut it out
EOD	End of discussion
GAL	Get a life
GJL	Go jump in the lake
GL	Get lost
GOOML	Get out of my life
GOWI	Get on with it
HHOS	Ha ha, only serious!
ITSTBF?	Is that supposed to be funny?
ITYLE	It took you long enough
KISS	Keep it simple, stupid
KMA	Kiss my ass

LCS	Last chance saloon
LMA	Leave me alone
LTBF	Learn to be funny
MA	My ass!
MYOB	Mind your own business
NOYB	None of your business
NRN	No reply necessary
NRNOE	No reply necessary, or expected
OTL	Out to lunch
OMG	OH MY GOD!
ONNTA	Oh no, not this again
PITA	Pain in the ass
QI	Quit it!
RB?	Reality bites?
RFD	Request for discussion
RUI?	Are you insane?
SOHF	Sense of humor failure
SLS	SO last season
SU	Shut up!

SWYP?	So what's your problem?
THS	Take a hike, sister
TAH	Take a hint
TMI	Too much information
TTC	That's total crap
T2TH	Talk to the hand ...
TWYT	That's what you think
TYLE	Took you long enough
WTH	What the hell
WYP?	What's your point?
WYSOH?	Where's your sense of humor?
XCM	Excuuuuuuuuse me!
YPB?	Your point being?
YGTBK	You've got to be kidding
YGTF	You've gone too far
YKIR	You know I'm right
YHL	You have lost
YHBW	You have been warned
YOYO	You're on your own

say it with symbols

They say that a picture paints a thousand words. And sometimes a suitably disgruntled gesture can say more than any text message could ever hope to achieve.

:-e	Very disappointed
(:-/	Very sad
:*(	In tears
>:-<	Mad
>:-<<	REALLY mad
@%&$%*******!!	(expletives deleted)
%-6	Braindead
(:-...	Brokenhearted
(|-(	Good grief!
8^(	Displeased
:-C	VERY displeased

:-(*)	You make me sick!
('^)	Looking away
:-p	Sticks out tongue
:-8(	Condescending stare
::-	THE FINGER
:-S	Words fail me
:-~<	Speak with forked tongue
:-~<<<<<	... VERY forked tongue
:-.	No comment
:= |	Stupid baboon
.:-|	Pea brain
3 o	Asshole
:-| :-|	Two-faced

OtOf100000SprmUHd2BTFstst

Out of a hundred thousand sperm, you had to be the fastest!

wish I hadn't said that

Sometimes a fight is good for clearing the air,

but the best part comes with making up.

[(xxx)]	A load of kisses
UWan2Tlk?	Do you want to talk?
PCB	Please call back
PCM	Please call me
NAH	Never again, honest
TMB	Text me back
UDNA	You didn't answer
IAU	I apologize unreservedly
IICO1OMLAIF?	If I cut off one of my legs am I forgiven?
LNFA	Let's not fight again

RNA	Ring no answer
WSLS	Win some, lose some
IMstCU	I must see you
BMP	Believe me, please
IMsdU	I missed you
IWSW	I was SO wrong
?:-(	What happened?
:-!	Foot in mouth
:+(	Hurt
ICBW	I could be wrong
ITBE	I take back everything
My#1	My number one
URT1	You are the one
MstBLuv	Must be love
(()):**	Hugs and kisses
K&MU?	Kiss and make up?
LtsT1k	Let's talk

CD42?

Candle-lit dinner for two?

IDdntMnIt

I didn't mean it

ILMkItUp2U

I'll make it up to you

HldMeClse

Hold me close

ImAFL4YrLuv

I'm a fool for your love

IDdntMn2HrtU

I didn't mean to hurt you

IMstvBnCrzy

I must have been crazy

LstW/oYrLuv

Lost without your love

LtsNtFiteAgn

Let's not fight again

URTLuvOvMyLif

You are the love of my life

ULiteUpMyLif

You light up my life

URTSnshnOfMyLif

You are the sunshine of my life

URALThtMTrs

You are all that matters

ILCUNMyDrms

I'll see you in my dreams

ILNvrLtUGo

I'll never let you go

ILuvUMrThnWrds CnSy

I love you more than words can say

NEdYrLuvSoBd

Need your love so bad

LtsNvrPrt

Let's never part

IJstKEpThnknAboutUBaB

I just keep thinking about you, baby

IWLAlwysLuvU

I will always love you

IWntU4evr

I want you forever

LtsSty2gthr

Let's stay together

U4Me4evr

You for me forever

TGILuvWU

This guy's/girl's in love with you

MIFrgvn?OKDrpM

Am I forgiven? OK, drop 'em!

:-) + (-: 4evr

You and me forever

Different strokes

A look at some alternative approaches to texting.

This is a different set of emoticons. These ones show their faces full-on – as if you were staring right at them. They don't need to be turned at 90 degrees.

^_^	Happy
;_;	Teary-eyed
(_o_)	Bowing
;^_^;	Embarrassed/out in a sweat
\=o-o=/	Wearing glasses
@_@	Bug-eyed
*^_^*	Blushing
-.-	Sleepy

-,-	Also sleepy
,,,^..^,,,	Cat peeking over a wall
mO-O m	Man peeking over a wall
moo?	Captain Hook peeking over wall
----	Invisible man peeking over wall
(^.^)/	Hi!
(@-@)	Stunned
(^o^)	Waaaaaaahhhhhhh!
(^_^)/-~	Smoking
(a-a)	Are you SERIOUS?
(o)(-)	Winker
(u_u)	Fast asleep
(+_+)	Cross-eyed
/\o/\	Spider
/\oo/\	Bat
/^_^\	Girl with pigtails
/\!/\	Just relaxing (male)
/*/\	Just relaxing (female)

\(^_^)/ Jump for joy

\O/ Praise the Lord!

=^.^= Cat

=^_^= Cat that got the cream

m(_ _)m Deep bow (Japanese style)

^V^ Ant eater

(^V^) Snowman with carrot nose

<(^o^)> Laughing loudly

(OvO) Hoot, hoot!

($_$) Rich kid

(*_*) The eyes of love

{{(>_<)}} Frrreeeeezing cold

<*_*> Alien

($_$) Richie Rich

[^_^] Geek boy

(>_<) Very angry

|_(..)_| Cyborg

(–_–) No emotions

(v_v)	Eyes popping out
(o-c)	Broken glasses
]^o^[	Cyberman
\8-8/	Bi-focal glasses
\~/	Full glass
_/	Empty glass
(O O)	Full moon
@(*O*)@	Koala

HOT TIPS

Making up your own little pictures can be great fun, but some people prefer to take a simpler approach to text messages. Instead of using emoticons, they just use a single letter to describe what they are feeling.

<G>	Grinning	<J>	Joking
<L>	Laughing	<O>	Shouting
<S>	Smiling	<W>	Winking
<Y>	Yawning	<Z>	Sleeping

A sideways glance

Here is another selection of emoticons. This time, for them to work, you need to imagine them as facial profiles.

o'!	Grim
o''	Pursed lips
o'j	Smile
o'P	Tongue out
o'r	Tongue out
o'T	Straight face
o'U	Yawn
o'V	Shout
o'Y	Whistle
o'\	Frown
o'v	Talking/happy
o'w	Forked tongue
o'\/	Daffy Duck

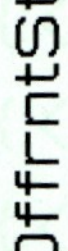

Guess the face

Here's an entertaining game for you play. All of the emoticons over the next few pages are of famous or identifiable characters. All you have to do is guess who they are.

#:O+ =	Betty Boop
((:=)X	Charlie Chaplin
{:<>	Daffy Duck
C8<]	Darth Vadar
:-) 8	Dolly Parton
:-)||?	Dr Who
5:-)	Elvis Presley
8)===;	Roadrunner
[:=|]	Frankenstein's monster
7:-)	Fred Flintstone
(:-)	Captain Picard

?:^[] Jim Carrey

8(:-) Mickey Mouse

:---) Pinocchio

(||) F RoboCop

*<|:-) Santa Claus

3 :-) Bart Simpson

(_8(|) Homer Simpson

@@@@:-) Marge Simpson

{8-* Maggie Simpson

{8-) Lisa Simpson

###O### Scary Spice

>:-| Mr Spock

(:=< Stormtrooper

<|-)= Fu Manchu

3:*> Rudolf

8:-[] King Kong

:O) Stimpy

=|:-)= Uncle Sam

=)||) Zorro

===:-D Don King

:_(Van Gogh

||||8^)X The Cat in the Hat

M-) See no evil

:X) Hear no evil

:-M Speak no evil

]B-) Batman

(P-| Borg

>>-O-> General Custer

HOT TIPS

Need some more flowers to send to your love one? Here is a little selection.

@}>-'-,--- %}------ @->->->-

(o)}->->- *U*-------| >>>>]E

@>+-+-+ -X-X-X--[] @@@----

Getting silly

For this game, the object is to get as silly, surreal, or oblique as possible. These should be tough for anyone to guess.

K%*} Drunk with lampshade on his head

G:-) NFL player

G-(Scuba diver with broken mask

K-[Squinting Batman

M:-) Military salute

O8-) Angel wearing shades

O>--< Dead guy on the road

Q=:-) Master chef

[O-| Cyclops wearing a Walkman

\.^./ Lotus position

\:-) Man in a beret

]=8) Happy cow

]B=8} Dragon

&&& Rubber chickens

<:-D Witch

<:-EXB Vampirella

<:>== Turkey

<:<)} Old hippie

=-O USS Enterprise

=-O~~~ Enterprise firing phasers

=====:} Snake

=|:-)= Abraham Lincoln

=|:-{|## ZZ Top

o(:-) Miner

:-)<//////> Bad tie day

(-o-) Imperial fighter

:-)>=> Bookworm

:>-< Stick 'em up

pp# Cow (profile)

q:-#	Baseball catcher
q:-)	Beastie Boy
_:^)	Native American
_o-)	Snorkeling
(: (	Ghost
(:::[]:::)	Band aid
(O--<	Fishbone
>>>>:===	Asparagus
%-^)	Picasso
(###]	Hockey player
4:-)	George Washington
8-#	The Grim Reaper
:-)j	Surfer
:-)K-	Shirt and tie
:-<	Walrus

:-E] I need to see a dentist

:-[#] Wears braces

s
:-|8 ( ) Pregnant woman

@ O=E<= Woman wearing turtleneck

c):-) Cowboy

c):-(=== Vomiting cowboy

c):-|* Sheriff

=={:-o] Genie in a bottle

|:-) Flat top

_/_/\o Caterpillar

(: = Beaver

=_= Getting sleepy

>[ ] Watching TV

!*|:-() Hit by a baseball bat

*-=|8-D Clown

Time to waste?

Finally, if you really have plenty of time on your hands, you can always try coming up with some of your own pictures.

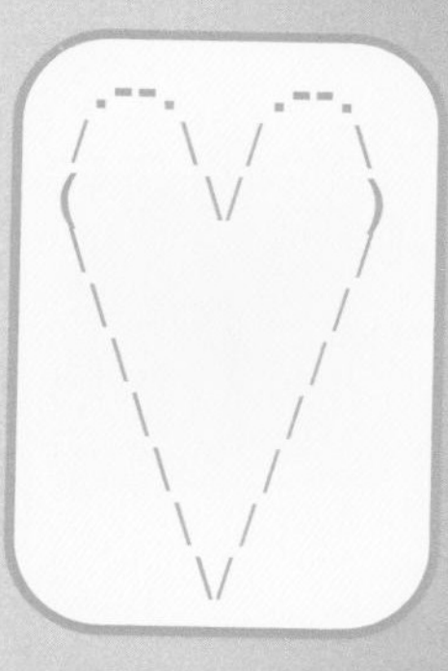

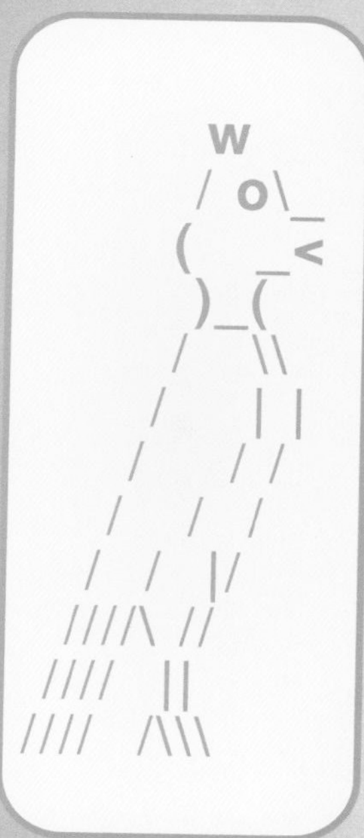

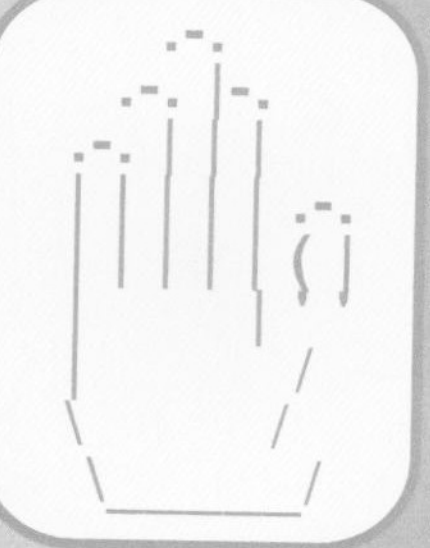